ALL ABOUT FOOD

VEGETABLES & HERBS

Cecilia Fitzsimons

Silver Burdett Press
Parsippany, New Jersey

First American publication
1997 by Silver Burdett Press
A Division of Simon & Schuster
299 Jefferson Road,
Parsippany, NJ 07054-0480

A ZOë BOOK

Original text © 1996 Cecilia Fitzsimons
© 1996 Zoë Books Limited

Produced by
Zoë Books Limited
15 Worthy Lane
Winchester
Hampshire SO23 7AB
England

Editors: Kath Davies, Imogen Dawson
Design: Sterling Associates
Illustrations: Cecilia Fitzsimons
Production: Grahame Griffiths

First published in Great Britain in 1996 by
Zoë Books Limited
15 Worthy Lane
Winchester
Hampshire SO23 7AB

Printed in Belgium by Proost N.V.
1 2 3 4 5 6 7 8 9 10

Library of Congress Cataloging-in-Publication Data

Fitzsimons, Cecilia
Vegetables & herbs/Cecilia Fitzsimons.
    p. cm.—(All about food)
"A Zoë book"
Includes index.
    Summary: Describes a variety of vegetables and herbs with tips on how to prepare them.
    ISBN 0-382-39593-X (LSB) ISBN 0-382-39598-0 (PBK)
    1. Cookery (Vegetables)—Juvenile literature. 2. Vegetables—Juvenile literature. 3. Cookery (Herbs)—Juvenile literature.
4. Herbs—Juvenile literature. [1. Cookery—Vegetables.
2. Vegetables. 3. Cookery—Herbs. 4. Herbs.] I. Title. II. Series.
TX801.F53 1997   95-26535
641.6'5—dc 20   CIP
                                                AC

Contents

Introduction

Thousands of years ago people wandered from place to place in search of food. They hunted wild animals, and they gathered wild plants to eat. The wild roots, leaves, and fruit provided most of their daily food.

When people began to settle in one place, instead of moving around, they found out how to grow wheat, beans, and vegetable crops.

People also discovered ways of keeping, or **preserving**, vegetables. They learned to store some crops in cool places and to dry beans and other vegetables. They dried some plants as herbs, which they used to flavor food, as we do today.

Now vegetables are also canned, bottled, and frozen to preserve them.

a clamp was used to store root vegetables outside

Facts about vegetables

The southern half of the earth is called the Southern **Hemisphere**. It is summer there when it is winter in the Northern Hemisphere.

When vegetable plants have stopped growing, or go out of season, in the Northern Hemisphere, they come into season in the Southern Hemisphere.

Fresh vegetables are shipped and flown across the world. They are in the shops all year round.

What is a vegetable?

A vegetable is any part of a plant that can be eaten. We eat leaves, stalks, **buds**, roots, flowers, fruit, and seedlings as different types of vegetables.

Grow your own vegetables

Many vegetable plants can be grown at home in the garden or in a pot or planting box on a balcony or windowsill.

On the next page are some tips to help you to grow healthy plants. All plants need soil, water, and sunlight for strong, healthy growth.

Sometimes the soil needs **compost**, a rich blend of decayed plant material, added to it to feed the plants. You can buy compost at a garden center, or you can make your own.

In the kitchen

You will find easy-to-follow recipes for different vegetables in this book. Here are some points to remember when you prepare food:

1. Sharp knives, hot liquids, and pans are dangerous. *Always ask an adult* to help you when you are preparing or cooking food in the kitchen.

2. Before you start, put on an apron and wash your hands.

3. All the ingredients and equipment are listed at the beginning of each recipe. Make sure that you have everything you need before you start.

4. Read the instructions. Measure or weigh the ingredients very carefully.

Think green

We often throw away things that we could use again, or **recycle**. If we reused some of our newspapers, cans, bottles, and plastic packaging, we would help to improve our **environment**.

Many vegetables are wrapped in packages when they are sold. When you read this book, you will find some ideas for things that you can do and make, using vegetables and their packages.

Planting in a pot

1. Take a clean plant pot. Place a few small stones over the holes in the bottom of the pot. This helps water to drain through the pot.

2. Fill the pot with soil enriched with compost. Make a hole in the middle. Gently put in your plant or seed. Take care not to damage a plant's roots.

3. Push the soil down around the roots or seed. Add more soil and press it down firmly.

4. Water the soil well and allow it to drain.

5. Place the pot in a saucer, dish, or pot holder to catch any water. Set it on a sunny windowsill or other well-lighted place. If planting seeds, place the pot in a plastic bag until the seedlings first appear.

6. Water regularly—once a week is usually enough. Occasionally feed with plant food, bought from a garden center. Follow the instructions on the bottle or packet.

Planting in the garden

You may need to **transplant** your pot-grown plants into the garden.

1. Dig a hole and put some well-rotted compost in the bottom of it.

2. Tap the bottom of your plant pot first to loosen the plant's roots. Remove the plant from its pot.

3. Gently place the plant in the hole. Replace the soil around the roots and press down firmly with your foot. Water and feed the plant regularly.

Seeds

You can buy seeds from shops, garden centers and mail-order catalogs. Follow the instructions printed on each packet. Each type of vegetable needs to be grown in a different way.

Seeds from fresh vegetables

You can also collect seeds from the fruits and pods of fresh vegetables. This often works well, although some seeds may not grow, or **germinate**. Try beans, peas, squash, and pumpkins.

Collecting seeds

1. Simply remove ripe seeds from the pods and plant them.

2. If the seeds are in a soft, sticky fruit flesh, or **pulp**, wash them in a strainer to remove the pulp.

3. Dry the seeds with a kitchen towel before planting.

4. Dried seeds can often be stored in an airtight jar for months.

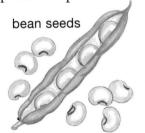

bean seeds

water

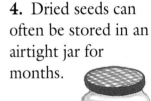

Think green

Grow seeds and plants in margarine tubs and yogurt containers. Make a hole in the bottom to let the water drain through.

Place clear plastic boxes or bottles (carefully cut off the bottom first) over each young plant to make a mini-greenhouse. This will keep slugs away in the garden.

If you have a garden, you can make your own compost. Put all your vegetable and garden waste in a pile in the garden. It will rot down into compost, used to enrich the soil.

Potatoes

The potatoes that we eat are the underground food-storing parts of the potato plant. These parts are called **tubers**. Tubers are full of **starch**, which is food to help the plant to survive the cold winter months. If left in the ground, each potato will grow into a new plant.

Potatoes are related to tomatoes. Potato plants have weak stems, divided leaves, and white or purplish flowers. There are hundreds of different types, or **varieties**, of potatoes.

Potatoes can be cooked in many ways— boiled, steamed, fried as potato chips and French fries, baked, or roasted.

Vast quantities of potatoes are canned, frozen, made into potato chips, dried mashed potatoes and other food products.

potato

More tubers

The Jerusalem artichoke is a type of sunflower from North America. It has crunchy, knobbly tubers that taste like globe artichokes.

Jerusalem artichoke

sweet potato

Sweet potatoes grow beneath a trailing vine. They are cooked as both a savory or sweet vegetable.

Yams are climbing plants that grow well in hot, wet, **tropical** countries. Some yams have enormous tubers that can grow up to about 3 feet (1 meter) long.

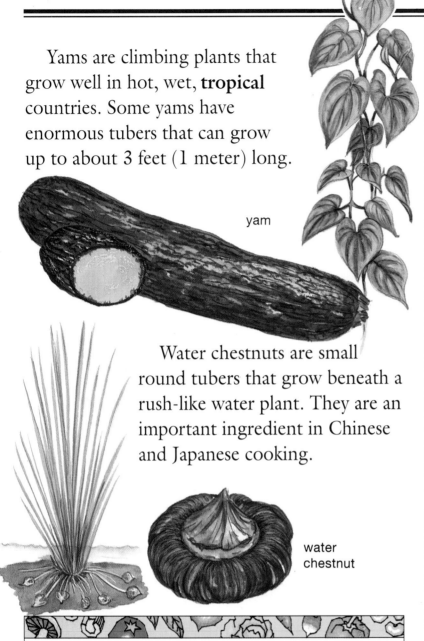

yam

Water chestnuts are small round tubers that grow beneath a rush-like water plant. They are an important ingredient in Chinese and Japanese cooking.

water chestnut

Facts about vegetables

Potatoes have been grown in South America for at least 2,000 years. They were brought to Europe by the Spanish about 400 years ago.

Potatoes are rich in **carbohydrates**, which our bodies need to produce heat and energy.

Jerusalem artichokes are not really artichokes, but are instead related to sunflowers.

"Fufu" is a thick, soupy, soft food made in Africa, using yams.

Making potato prints

You will need:

1 potato

liquid poster paint

a shallow dish

paper

a felt-tip pen

a sharp knife

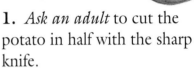

1. *Ask an adult* to cut the potato in half with the sharp knife.

2. Use a felt-tip pen to draw a simple design on the cut surface of the potato.

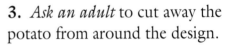

3. *Ask an adult* to cut away the potato from around the design.

4. Dip the potato, cut side down, into the paint.

5. Press the potato onto the paper.

6. Gently lift the potato up to see your print. Repeat to make a pattern.

Root vegetables

Some plants, such as beets, store their food in one swollen root. Other plants, such as carrots, have long, straight **tap roots** that reach deep into the ground.

We cook most root vegetables before we eat them. They can be boiled, steamed, baked, or roasted. Many root vegetables, such as carrots and radishes, are eaten raw in salads.

Carrots and parsnips have umbrella-shaped clusters of flowers and finely divided leaves. They are related to celeriac, which has a thick short root and is a form of celery.

parsnip

carrot

celeriac

flowers

turnip

rutabaga

Several members of the cabbage family produce swollen roots—such as turnips and rutabagas. The plants have large leaves and spikes or clusters of 4-petaled flowers. Rows of seed pods contain large numbers of tiny round seeds.

Radishes are related plants with smaller divided leaves.

radish

beet

Beets are descended from wild Sea Beets, which still grow on beaches today.

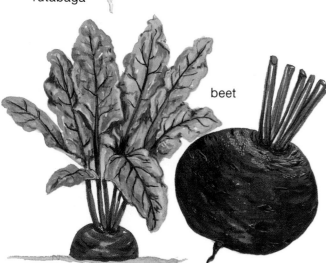

Cassava is one of the most important food plants in tropical countries. The root contains starch. It grows beneath a **shrub** about 6.5 feet (2 meters) tall. The root is dried and then ground. People make a type of flour called manioc from cassava roots.

cassava

Facts about vegetables

Carrots, parsnips, and beets contain sugar. Sometimes they have been used to sweeten cakes and other desserts.

Children need **vitamin** A to help them grow. Carrots are full of carotene, which is a source of vitamin A.

Radishes were eaten in Egypt more than 3,000 years ago.

Mooli is a huge radish from Japan that can grow up to about 3 feet (1 meter) long. It has a milder taste than other radishes.

Rutabagas are large yellow turnips. They were brought to Britain from Sweden. The British call them swedes.

Carrot tops

You will need:

1 carrot

a sharp knife

kitchen towel

an empty plastic yogurt or margarine container

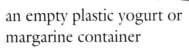

1. *Ask an adult* to cut a slice from the top of the carrot, below the base of the old stalk, using a sharp knife.

2. Moisten a piece of kitchen towel with water, and place it in the bottom of a plastic container.

3. Place the piece of carrot, cut side down, on top of the wet paper.

4. Set the container on a sunny window ledge. Water occasionally, and you will soon see new leaves begin to grow.

Draw a funny face on a sticky label and fix it to the side of the container. As the leaves grow, they will look like hair!

Try using different types of vegetable tops. See if they all grow the same-shaped leaves.

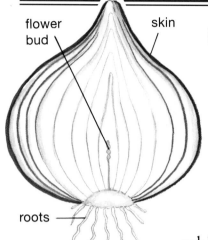

flower bud

skin

roots

Onions, garlic, and leeks

Onions are related to lilies and to daffodils. Each plant grows from layers of fleshy leaf bases that surround a tiny flower bud. This is called a **bulb**. The bulb is covered by an outer layer of dry, papery skin. Round clusters of tiny flowers are carried on a single long stalk. Onion flowers are greenish white. The leaves are long and hollow. There are over 500 varieties of onions. The bulbs can be many different shapes, sizes, and colors. Most onions are white, but some are red, yellow, or purple.

onion flower

Onions can be eaten raw or boiled, baked or fried. They are used to make pickles, **chutneys**, and sauces and to flavor many savory dishes. Onions are also frozen, dried, or made into powder for food flavoring.

onion

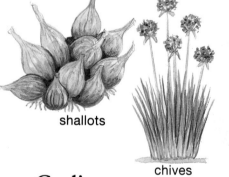

shallots

chives

spring onions

Shallots are small onions that grow together in a cluster. Chives are small grasslike onions. They are used as an herb to flavor foods. Spring onions are picked young and are eaten raw in salads.

garlic

Garlic

Garlic grows best in warm, dry areas. It has small white flowers and long flat leaves. Each garlic bulb is made up of a cluster of smaller bulbs, or **cloves**.

Garlic is widely used to flavor food. It can be eaten raw or cooked.

12

Leeks

Leeks are bulbs with tall stalks, long flat leaves, and purple flowers. They are cooked as a vegetable or made into soups, flans, and pies. Leeks can be frozen, dried, canned, or bottled.

leek

Facts about vegetables

Garlic is good for you! It is an **antiseptic**, which means that it kills germs.

Onions probably first came from central Asia and were grown in China and Egypt more than 3,000 years ago.

Christopher Columbus introduced the first onions to North America about 500 years ago.

Welsh people wear leeks on St. David's Day, March 1. In A.D. 640, Welsh soldiers won a battle against the Saxons because the Welsh were able to recognize one another. They all wore leeks in their hats.

Onion dip

You will need:

1 small onion

1 small container of sour cream or yogurt

1 pack of plain potato chips

a sharp knife

a spoon

a bowl

1. *Ask an adult* to peel and then chop up the onion very finely with a sharp knife.

2. Put the chopped onion into the bowl.

3. Pour the sour cream or yogurt into the bowl.

4. Stir with a spoon until the onions are mixed in well.

5. Put the mixture in the refrigerator for a couple of hours to chill.

6. Serve as a dip with potato chips.

Shoots and stems

celery flower

celery

In some plants it is the stem or young shoots that swell and can be eaten.

Celery and fennel are both relatives of carrots and parsnips. They have long stalks, divided leaves, and an umbrella of flowers. Celery is eaten raw in salads, cooked as a vegetable, or used to flavor stocks and soups.

Fennel is a short plant whose leaf bases are swollen to form a type of bulb, which is cooked or eaten raw in salads.

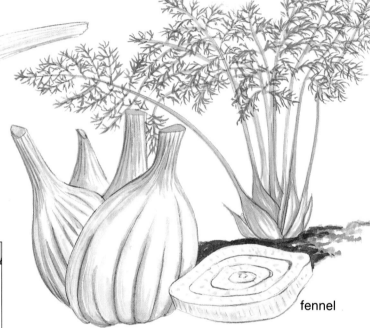

fennel

Facts about vegetables

Celery is good for your teeth! It has the mineral calcium in it. Our bodies need calcium to develop strong bones and teeth.

Celery was used as a medicine for thousands of years. It was first eaten as a vegetable about 300 years ago.

In ancient times it was thought that fennel had the power to give strength and courage.

Asparagus first came from Eastern Europe. It was grown by the ancient Egyptians, Greeks, and Romans.

Chard (silverbeet) is related to the spinach beet. The leaves taste like spinach. Thick chard stalks are cooked as a separate vegetable.

chard

Asparagus is a type of lily that grows in sandy soils. In spring thick shoots grow up from the roots and are cut when they are about 6 inches (15 cm) long. These are called asparagus spears. If left, they will grow into tall stems with fernlike leaves, tiny flowers, and red berries.

Asparagus is cooked and is eaten with melted butter. It can be canned, bottled, or frozen and is used in soups and flans.

asparagus

Bamboo shoots from tropical Asia are used in Chinese, Japanese, and Korean cooking. Bamboo is a type of grass, and only the youngest shoots can be eaten as they grow up from the roots. Pieces of shoots are cut when they are 6 to 12 inches (15 to 30 cm) long. Bamboo shoots are canned, bottled, or pickled.

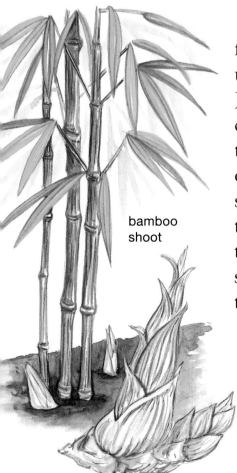

bamboo shoot

Cheese and celery boats

You will need:
fresh celery
1 container of cream cheese
1 pack of plain potato chips

a sharp knife
a spoon
a serving dish or plate

1. Carefully remove a few celery stalks and wash them.

2. *Ask an adult* to trim the ends and cut the celery stalks into 3 inch (7cm) lengths.

3. Use a spoon to fill the celery pieces with cream cheese.

4. Put the celery "boats" on the serving dish or plate.

5. Stick a potato chip in each boat to make a sail.

Peas and pods

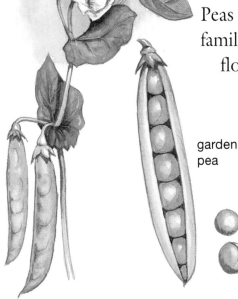

Peas and beans are part of one of the largest plant families, the **legumes**. They all have butterfly-shaped flowers and carry their seeds in pods. The seeds are called **pulses**.

garden pea

Peas are climbing plants, which grow up to 5 feet (1.5 meters) tall. They have white, scented flowers. The pea pod dries when it is ripe and springs open to scatter the seeds.

There are many varieties of peas. Sugar, or snow, peas have sweet, tender pods. The pods are cooked and eaten whole, with the peas inside them.

Petits pois are tiny peas that come from France. Marrowfat peas have large, floury seeds, which are dried or canned. Garden peas are smaller. Fresh peas can be eaten raw or cooked. Peas are also canned, bottled, or frozen.

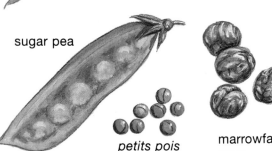

sugar pea

petits pois

marrowfat peas

Lentil plants have small leaves and white flowers. The short, flat pods contain 1 or 2 seeds. Green, brown, or red lentils are sold dried. They can be cooked as a vegetable or used to thicken soups and stews.

lentils

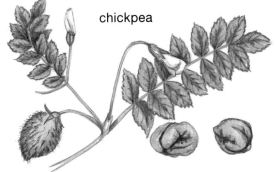

chickpea

Chickpeas are one of the most important foods grown in India. They have feathery, divided leaves and small hairy pods containing 1 or 2 seeds. The yellow or black seeds have a mild, nutty flavor. Chickpeas are sold dried or canned.

Cowpeas are sometimes called "yardlong beans" because their pods can grow up to 1 yard (1 meter) in length. They were first grown in Africa.

Young, fresh pods are chopped and eaten like green beans. Dried peas, called black-eyed peas, are used in soups and stews. They can be ground into flour or made into a paste that is used in many dishes such as hummus.

cowpeas

Asparagus peas are eaten whole, like sugar peas. Each pod has several leafy wings along its length.

asparagus pea

Facts about vegetables

Pulses are rich in **protein**, which our bodies need for growth and to stay healthy.

Peas were grown in Myanmar more than 10,000 years ago.

A lens is the same half-moon shape as a lentil. That is why the word *lens* was given to this invention.

Hummus

You will need:

1 can of chickpeas

2 tablespoons of lemon juice

3 cloves of garlic

Slices of pita bread

a large strainer

a tablespoon

a large mixing bowl

a serving plate

1. Use the strainer to drain the water from the chickpeas and put them in the mixing bowl.

2. Mash up the chickpeas with the tablespoon, until they make a thick paste.

3. *Ask an adult* to peel the cloves of garlic and chop them finely with a sharp knife.

4. Add the chopped garlic and the lemon juice to the paste and mix them in well.

5. Spoon the mixture into a serving dish.

6. Serve the hummus with slices of pita bread.

Beans

broad bean

Beans, like peas, are part of one of the largest plant families—the legumes. There are many varieties of beans, all rich in protein.

Broad bean plants have straight, upright stems and black and white flowers. The beans are "shelled" from their fur-lined pods. Then they are cooked, canned, dried, or frozen.

scarlet runner beans

French bean

Scarlet runner beans are also called stick beans. These plants are vigorous climbers and are trained to grow up canes. They have bright red flowers.

French beans have short, rounded pods and grow on dwarf, bushy plants, or climbers. Most have white flowers, but some have purple flowers and produce purple beans.

Scarlet runner beans and French beans are cooked whole and eaten fresh, canned, or frozen.

haricot beans

Haricot beans are French beans grown for their dry seeds. They are used to make baked beans and other dishes.

butter beans

Butter beans are grown in many warm countries. They look like broad beans, but they grow larger seeds. Lima beans are similar to butter beans.

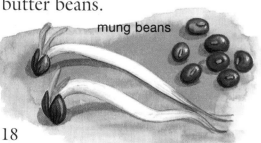

mung beans

Mung beans are small seeds grown as bean sprouts (bean shoots). They are an important ingredient in Chinese and Indian cooking.

red kidney beans

Red kidney beans, come from South America. They are also called chili beans because they are often cooked with chilies to make hot, spicy dishes.

soybeans

Soybeans are tiny, but they contain large quantities of protein and oil. These are extracted and used in many kinds of processed foods.

Facts about vegetables

Never eat raw broad beans. They contain **toxins** that are poisonous. These are destroyed when the beans are cooked.

Scarlet runner beans and French beans were first grown in South America. They were brought to Europe about 400 years ago.

Soybeans can be processed into **synthetic** meat, called Textured Soy Protein (TSP). If soybeans were planted on the land used to raise animals, they would produce more than 10 times as much protein as the meat from these animals.

Grow your own bean sprouts

You will need:

dried mung beans

a large plastic yogurt container with a tight-fitting lid

1. Soak half a cupful of beans in water overnight.

2. Put the beans in the bottom of the yogurt container.

3. *Ask an adult* to make some holes in the lid. Put the lid on the container, and place in a warm, dark cupboard.

4. Every morning and evening, fill the container with water and replace the lid tightly. Tip the container upside down and drain the water out over a sink. Replace the beans in the dark cupboard.

5. Your bean sprouts will be ready to eat in 3 or 4 days' time.

Leaves, buds, and flowers

The leaves of many plants can be eaten safely. They are **edible** plants. Most leafy vegetables are steamed or boiled. Many are canned or frozen. Some are eaten raw in salads or used to make pickles.

Cabbages grow as a large bud of tightly-packed leaves. They have thick, leathery leaves and a yellow spike of flowers. Some cabbages have white smooth leaves; others have green rough, wrinkled leaves. Red cabbage has red or purple leaves.

Brussels sprouts are small cabbage-like buds that grow at the base of each leaf. Sprouting broccoli are loose clusters of green, white, or purple flower stalks.

cabbage

Brussels sprouts

sprouting broccoli

cabbage flowers

cauliflower

Cauliflower is a large cabbage-like flower head with white **florets**. All these plants are members of the cabbage family.

Spinach is a loose-leafed plant that is related to beets. Its long green leaves turn dark green when they are cooked.

spinach

Chinese cabbage

Chinese cabbage is a vegetable that looks like a tall lettuce head. There are several varieties, which can be eaten raw or cooked.

globe
artichoke

Globe artichoke is one of the few flowers that we eat. Artichoke plants grow up to 6 feet (2 meters) tall and have greyish leaves. We cook and eat the young flower buds. If they are left to grow, a beautiful blue thistle flower blooms.

Facts about vegetables

All the plants in the cabbage family are descended from the wild cabbage, which grows in many areas.

Green vegetables are rich in vitamin A. Spinach also contains iron. We need iron to keep our blood healthy.

Broccoli was first grown in Italy. *Broccoli* is an Italian word that also means a small shoot or sprout.

Coleslaw

Coleslaw is a Dutch word which means cabbage salad.

You will need:
a small cabbage
1 small carrot
1 small onion
mayonnaise

a knife
a grater
a bowl

1. *Ask an adult* to slice the cabbage finely. Put the cabbage into a large bowl.

2. *Ask an adult* to chop the onion finely and to grate the carrot.

3. Put the chopped onion and grated carrot into the bowl with the cabbage.

4. Mix well and moisten with mayonnaise.

Leafy salads

Lettuces and other soft-leaved plants are usually eaten raw in salads.

Lettuces are members of the daisy family. They produce a spike of yellow, daisy-like flowers and parachute seeds like a dandelion.

There are many varieties of lettuce, which can be grouped into four types:

butterhead lettuce

Butterhead lettuces, such as Boston lettuce, have dense, round heads. Their leaves are soft and crunchy.

Head lettuces, such as Iceberg, have a dense head of thick, crispy leaves.

Romaine lettuces have tall, upright heads of long, narrow leaves.

Loose-leaf lettuces have loose, open heads of curly, toothed leaves. Some varieties are red.

Iceberg lettuce

lettuce flowers

romaine lettuce

loose-leaf lettuce

endive

Endives grow a wide head of tightly-packed, curly leaves. Their pale blue flowers grow on a tall flower stem. Endive leaves taste bitter. The plants are covered and left in the dark, or **blanched**, for 5 to 10 days before they are harvested. Chicory leaves are also blanched before harvesting. Both endives and chicory are often eaten raw, but they can also be cooked as a vegetable.

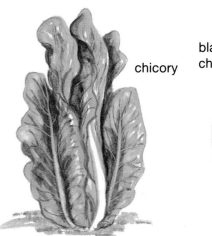

chicory

blanched chicory shoot

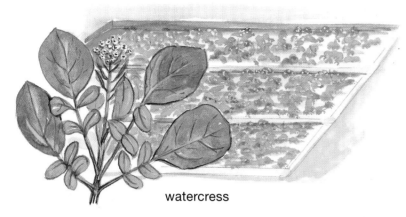

watercress

Watercress is a water plant. It is grown in watercress beds, which are flooded with clean, **unpolluted** water. Watercress is eaten raw or cooked in soups.

mustard and cress

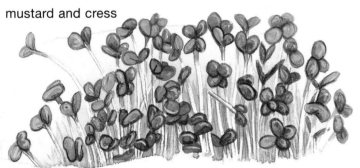

Mustard and cress are sprouted seedlings. Like watercress, they have a hot, peppery taste. They are used in salads and sandwiches.

Facts about vegetables

Lettuce, chicory, and endive were used as **medicinal herbs** for more than 1,000 years. Lettuces may make you sleepy!

Roasted chicory roots can be ground up and used instead of coffee.

Watercress is a natural, multivitamin health food. It contains vitamins A, B, C, and E, as well as iron and calcium.

Grow your own mustard and cress

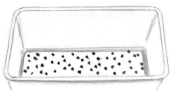

You will need:

mustard and cress seeds

a paper towel

a clean yogurt container or margarine container

a clean plastic bag

1. Line the bottom of the container with a folded piece of paper towel. Moisten with water.

2. Put mustard and cress seeds on the paper towel.

3. Cover the container with a plastic bag and set it in a dark place.

4. When the seedlings are 2 inches (5 cm) tall, place the container on a window ledge until the dark green leaves have formed.

You could add a few mustard and cress leaves to go with lettuce in a salad. Cress tastes good in egg salad sandwiches as well.

Cooking herbs

bay

Cooking herbs are sweet-smelling plants whose leaves and stalks contain scented oils. Dried or fresh herbs are added to many cooked foods to flavor them. Fresh herbs are also used raw in salads and to garnish food.

Bay is an evergreen tree that can grow up to 30 feet (9 meters) tall. The dark, leathery leaves are used to flavor soups and stews.

Many herbs belong to the mint family They all have spikes of two-lipped flowers that grow from the base of the leaves. Mint grows up to 3 feet (90 cm) tall. It is used to flavor vegetables and to make mint sauce, which is served with lamb.

mint

sage

rosemary

Sage is a shrub with woody stems, greyish leaves, and large purple flowers. Sage and onion stuffing is served with pork and duck.

Rosemary is a large shrub with narrow, dark green, needle-like leaves. It is often cooked with pork and lamb.

Fennel grows to about 4 feet (1.2 meters) and has fine, feathery leaves that taste of aniseed. Fennel leaves are used to flavor fish. Dill is similar in look and taste to fennel, but the plant does not grow as tall.

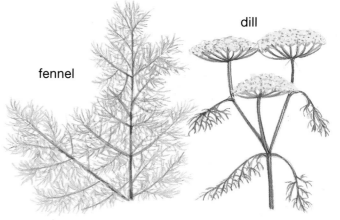

dill

fennel

parsley

Parsley grows to about 2 feet (60 cm) tall. Its leaves are divided into small leaflets and may be curled or flat. Parsley is added to many dishes and is used fresh in salads and as a garnish.

Basil grows best in warm, dry climates. It has bright green rounded leaves and small whitish flowers. The plant grows to about the same size as mint. Basil is used to make pesto, an Italian sauce that is eaten with pasta.

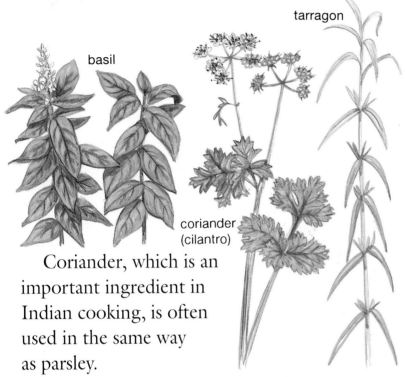

basil

tarragon

coriander (cilantro)

Coriander, which is an important ingredient in Indian cooking, is often used in the same way as parsley.

Tarragon is a shrub-like plant that grows about 3 feet (90 cm) tall. It has long, narrow green leaves and small green flowers. Tarragon has a strong flavor that goes well with chicken. It is often used to flavor oils and vinegars.

Marjoram, also called oregano, is a small shrub with small, dark green leaves. It is often added to tomato-flavored dishes.

Thyme is also a small shrub with short, broad-pointed leaves and small pink flowers. It looks a little like marjoram, but it has a different scent and a much stronger flavor. It is added to many dishes and is used fresh in salads.

Grow your own herbs

Some herbs will grow easily in pots kept inside or on the windowsill.

Parsley and mint are easy to grow from seed. You can grow basil from seed as well, but it needs a lot of sunshine and warmth to grow well.

Thyme, sage, rosemary and marjoram are easier to grow from seedlings.

(Look at pages 5 to 7 of this book for tips about growing healthy plants.)

Facts about herbs

Herbs have been grown for thousands of years. Once they were used mainly for medicines, rather than for flavouring foods. Many herbs are still used in medicines today.

The first tea drinks in Europe were herb teas, or **tisanes**.

Some of these plants also have flavored seeds, which are used in cakes or ground like spices.

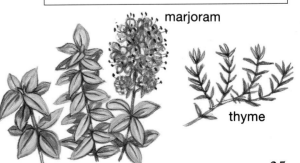

marjoram

thyme

Vegetable fruits

tomato

These "vegetables" are actually the fruits of their plants. They protect and feed the seeds that grow inside them. They do not taste as sweet as other fruits and as a result are eaten in the same way as vegetables.

Tomatoes, peppers, and eggplants are related to potatoes. They all have similar star-shaped flowers and shiny, berry-like fruits.

Tomato plants are straggly bushes with clusters of small yellow flowers. Most tomatoes are bright red, but some are yellow, pink, or white.

Tomatoes are eaten raw or cooked. They can be canned, bottled, dried, or made into a thick sauce, or purée, and tomato juice.

Sweet peppers have hollow fruits that are red, yellow, purple, or white when ripe, but green when unripe. Chili peppers are small pointed peppers that taste extremely hot. Chilies are an important part of Mexican cookery.

sweet pepper

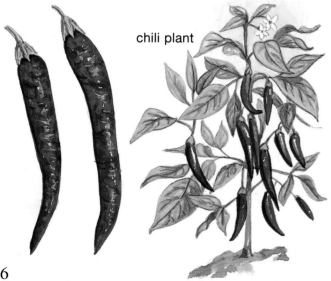

chili plant

Facts about vegetables

Tomatoes, peppers, chilies, and avocados were all first grown in South and Central America. They were unknown in Europe until about 500 years ago.

Sweet peppers contain large amounts of vitamin C and carotene.

Avocados are rich in oils and proteins. The oil is sometimes used to make soap.

Eggplants were first grown in China more than 1,500 years ago. The white or deep purple fruits can grow up to 12 inches (30 cm) long.

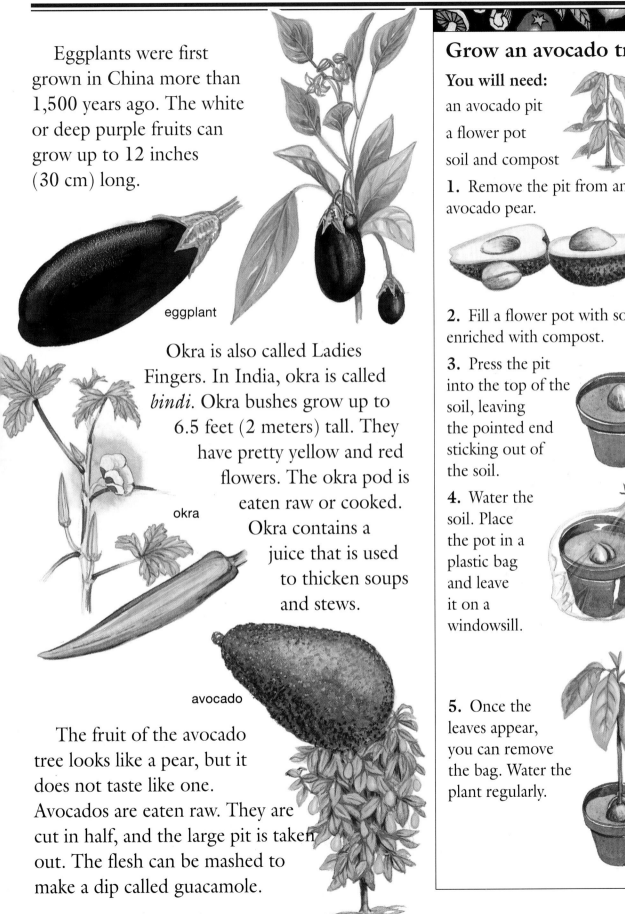

eggplant

Okra is also called Ladies Fingers. In India, okra is called *bindi*. Okra bushes grow up to 6.5 feet (2 meters) tall. They have pretty yellow and red flowers. The okra pod is eaten raw or cooked. Okra contains a juice that is used to thicken soups and stews.

okra

avocado

The fruit of the avocado tree looks like a pear, but it does not taste like one. Avocados are eaten raw. They are cut in half, and the large pit is taken out. The flesh can be mashed to make a dip called guacamole.

Grow an avocado tree

You will need:

an avocado pit

a flower pot

soil and compost

1. Remove the pit from an avocado pear.

2. Fill a flower pot with soil enriched with compost.

3. Press the pit into the top of the soil, leaving the pointed end sticking out of the soil.

4. Water the soil. Place the pot in a plastic bag and leave it on a windowsill.

5. Once the leaves appear, you can remove the bag. Water the plant regularly.

Pumpkins and mushrooms

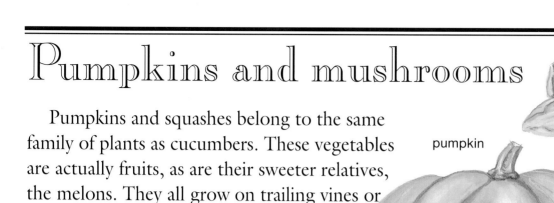

Pumpkins and squashes belong to the same family of plants as cucumbers. These vegetables are actually fruits, as are their sweeter relatives, the melons. They all grow on trailing vines or short, bushy plants and have large, hand-shaped leaves. The large, watery "fruits" grow beneath yellow flowers.

pumpkin

Pumpkins are some of the largest vegetables. Some pumpkins have reached about 500 lbs (225 kg). Squashes are similar to pumpkins. There are many shapes and varieties of squashes.

Zucchini are a variety of squash, picked when young. The flowers can be fried in batter.

squashes

zucchini

Pumpkins and squashes are boiled, fried, or baked. They can all be used to make soups, stews, pickles, jams, and casseroles.

Facts about vegetables

The inside of a dried dishcloth gourd is full of fibers that make a bath sponge or "loofah." Gourds are related to squashes and pumpkins.

The first greenhouse cucumbers were grown by the Romans for the Emperor Tiberius, who wanted to eat them all year round.

Pumpkin seeds contain fats and protein. *Pepitos* are fried and salted pumpkin seeds.

Cucumbers first came from southern Asia and were grown in India 3,000 years ago. Now cucumbers are either long greenhouse varieties or shorter ridged types grown outside. Gherkins are small cucumber-like fruits that are pickled in **brine**.

greenhouse cucumber

gherkin

ridge cucumber

Mushrooms

Mushrooms look like plants, but they have no green tissues to trap energy from the sun. Instead they obtain their food from other rotting plant and animal materials. They are **fungi**. Mushrooms are eaten raw, boiled, or fried and can be bottled, canned, dried, or frozen.

mushroom

WARNING:
Never pick or eat wild fungi, unless you are with an adult who knows about them. Some fungi are extremely poisonous.

Make your own shaker

You will need:

a squash or pumpkin
kitchen towel
a sharp knife
a tablespoon
a strainer

1. *Ask an adult* to cut the squash, or pumpkin in half, using a sharp knife.

2. Remove seeds with a spoon; put them in a strainer.

3. Wash the seeds to remove the pulp.

4. Put the seeds on kitchen towel to dry.

5. When the seeds are completely dry, put them in a plastic pot with a tight-fitting lid—and shake!

Glossary

antiseptic: a substance that can destroy germs that cause disease.

blanch: to make something white. Blanching makes some vegetables less bitter in taste.

brine: very salty water that is often used to pickle or preserve foods.

bud: the very young shoot or flower of a plant.

bulb: the underground bud of a plant in which food is stored.

carbohydrate: a substance made up of carbon, hydrogen, and oxygen, such as sugar, and starch, which is an important part of our diet.

chutney: a thick sauce of fruit and vegetables cooked together with spices, sugar, and vinegar to preserve them. Chutney was first made in India.

cloves: the divided parts of a bulb.

compost: a rich blend of decayed plant material, such a leaves, used to enrich the soil.

edible: something that is fit to be eaten.

environment: everything around us, such as air, water, and land.

florets: small flowers.

fungi: plants that have no leaves or flowers, and which grow on other plants or material that is living or dead.

germinate: when a seed starts to grow into a plant.

hemisphere: one half of the earth.

legumes: plants that produce fruits in a pod.

medicinal herbs: plants that are grown for use as medicines.

preserve: treating food so that it does not go bad.

protein: a substance that people and animals need to live and grow.

pulp: the juicy flesh of a fruit.

recycle: to make something new from something that has already been used.

shrub: a woody plant or bush that is like a small tree.

starch: a white, energy-rich food substance without taste or smell, which is in foods such as potatoes and bananas.

synthetic: a substance or material that is made by people and is not produced naturally.

tap roots: the strong main roots of plants that grow straight down into the soil.

tisanes: drinks made by boiling down dried herbs with water.

toxins: poisons that are not made by people but occur naturally.

transplant: to replant in another place.

tropical: from the lands near the middle of the earth, where the heat from the sun is strongest. We draw lines on maps to show the position of the tropics. Vegetables from the tropics grow best in hot, wet places.

tubers: swellings in the stems or roots of plants, where food is stored. Tubers are usually underground.

unpolluted: clean and not harmful to human, animal, or plant life.

variety: a type of plant or fruit.

vitamins: the small amounts of different substances in foods that people and animals need for good health. Most vegetables contain vitamins A and C.

Further reading

Food and Feasts: In Tudor Times by Richard Balkwill. New Discovery Books, 1995. This introduction to the Tudor Period examines the food the people ate and where it came from and includes authentic, delicious recipes.

Eat Well by Miriam Moss. Crestwood House, 1993. Young readers will learn how to plan a diet to help them stay healthy and will get practical advice that can help them improve their confidence and well-being.

Index of vegetables and herbs